Bibliographic information published by the German National Library:

The German National Library lists this publication in the National Bibliography; detailed bibliographic data are available on the Internet at http://dnb.dnb.de .

Imprint:

Print and binding: Books on Demand GmbH, Norderstedt Germany
ISBN: 9783668560086

This book at GRIN:

http://www.grin.com/en/e-book/375396/literature-review-on-the-performance-of-reduced-weight-tendon-tlp-in-ultradeep

Syed Ahmad Fathi Syed Mohd Khair

Literature Review On The Performance of Reduced Weight Tendon TLP in Ultradeep Water

GRIN Publishing

LITERATURE REVIEW ON THE PERFORMANCE OF REDUCED WEIGHT TENDON FOR TENSION LEG PLATFORM (TLP) IN ULTRA-DEEP WATER

Syed Ahmad Fathi bin Syed Mohd Khair

1.0 Abstract

The fast growing offshore industry has come up with advanced technology that makes it possible to produce oil and gas on deeper offshore sites. One of the interests now is to look for a structure and systems that are reliable and cost effective for ultra deep water operation. Tension leg platform (TLP) concept is currently been viewed for this ultra deepwater operation. TLP differ from other type of floating platform in a sense that the hull buoyancy exceed its weight, the hull is vertically moored at a draft below the displacement thus creating constant tension in the mooring line. The problem for the development of TLP beyond 1500 meters water depth is the material for the tendon. Conventional tendon made by steel is too heavy and proven to be unpractical in ultra deep water. New material is needed, that is lightweight and stiffer, and this material needs to meet the structural performance characteristic for practical development. This research is intended to demonstrate the performance of reduced weight tendons in ultra-deep water. The effect of weight reduction on motion response of the TLPs in regular and irregular waves and the performance differences between conventional tendons and reduced weight tendons in term of motion response also will be investigated. Based on the review it is known and widely accepted that numerical analysis is the most practical and economical way to analyze the structure, but because of the simplification and assumption of the numerical calculation a validation is needed. The numerical analysis will be done using industrial accepted software, ANSYS AQWA. For the purpose of validation, model test is proposed to be done in the test basin, but due to very limited water depth in the test basin, truncation method will be used instead of common traditional test. This method also known as hybrid method that uses numerical calculation to predict the cut off portion of the tether lines and using physical model test in the upper portion.

1.1 Introduction

The motion of a structure floating or submerged in the sea and exposed to various environmental loads have long became an interest of scientific and industrial communities all over the world. Understanding and predicting this motion is very essential for the design, development and safe operation in the sea especially in remote areas. The idea of using a compliant offshore platform such as TLP was not entirely a new idea, works have been done to develop this kind of system start already in sixties. This chapter of literature review will discuss a great deal of works that have been done considering various aspects that affect the design and safe operation of the TLP. These include the environmental loads that the structure will endure, the tendon integrity itself and the various type of analytical approach. Numerous studies also been reviewed that address various response of the TLP when exposed to the random loading. Finally the literature review will discuss the type of model test that have been done to address the ultra-deep water condition and an overview of rules and regulation for these compliant structure.

1.2 Environmental loads

The most important part that needed to be considered in any of the design, construction, installation, and operation of any offshore structure is the environment that the structure will be placed in. This environmental factor or forces that act on the structure involved waves, wind, tide, and current. Other extreme cases such as earthquake, hurricane and storm also sometimes seem realistic to be considered. Failure to understand or predict all these various forces spell catastrophe.

The basic major forces that act on the fixed offshore structure are wind, waves and current. Various studies until this day focus on these forces. Li and Kareem (1990) conducted a study of the dynamic behavior of the TLP under coincident loading of wave fields and random wind. To address and accurately evaluate the diffraction and radiation forces, a computer code that based on the boundary element method (BEM) is developed.

In the analysis they found an excellent agreement between the frequency domain and time domain analyses with the response of a typical TLP in six degrees of freedom. This agreement is manifested using power spectral density of the surge response of the structure.

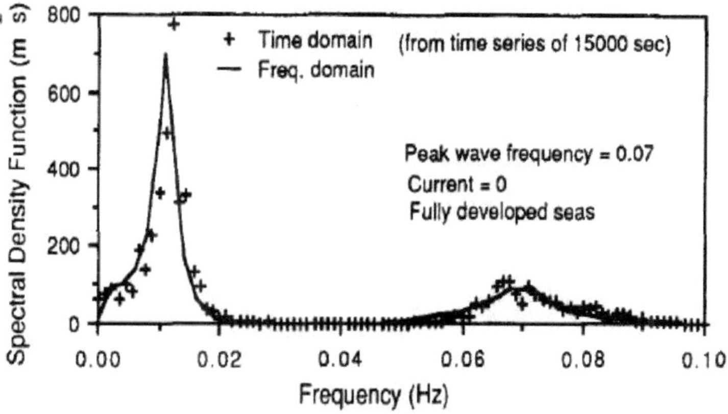

Figure: Power Spectral Density (PSD) of TLP Surge Response (Li, 1990)

The effect of regular waves on TLP is presented by Jain (1995). The nonlinear coupled response of the TLP is explained in detail base on the dynamic analysis. He noted that because of the natural periods in surge, sway and yaw is evidently more than heave, roll and pitch the resonance that correspond to surge, sway and yaw is not likely to take place. Jain also found that the resonance effect with heave frequency could trigger a large fluctuation in the tension of the TLP tendon. This happen primarily due to the fact that the heave period is regularly close to wave periods that frequently occur. With all of this good result especially in the TLP response with respect of surge, sway, yaw, heave, roll and pitch, it seems that more need to be done to understand the behavior of the TLP response, especially in regard of irregular wave which reflect a closer reality which was happening in the sea.

The fast growing industry also triggers the study of effect of the incoming waves and their effect to the TLP in greater detail. S. Chandrasekaran et al. (2006) presented a technical note discussing the response of TLP from varying waves angle. The Stokes fifth-order nonlinear wave theory is used to model the hydrodynamic loading. They concluded that the change in the angle of wave approach have a direct influence to the dynamic response of the triangular TLP. This is true to all degree of freedom except for the heave response. They also observed that the response of the TLP from pitch and roll point of view are highly stochastic in nature. This shown that these motions indicate high degree of randomness. These finding maybe true for TLP with triangular geometrical design, more need to be done for other type of TLP geometry model to further confirm all these responses.

Extreme wave effect on the TLPs is also an essential part that need be understood for the development of TLPs in deepwater. T.B. Johannessen et al. (2006) presented a study on extreme wave effects based on Snorre A model test. The study focus on accidental wave event and how to analyze them. It also touched on the measure need to be taken for the design. This study however may not applicable to the deepwater or ultra deepwater. The Snorre A TLP installed in the shallow water in the North Sea, with only 308 meters water depth.

To evaluate the tether loading accurately due to harsh environmental forces T.B. Johannessen et al. conducted hydrodynamic model test with a scale 1:62.5 in the wave basin. This test is expected to clarify some of the major concern such as offset and slowly varying motions, drift forces and direct response to waves. Although the test is a great help in clarifying these issue some constraint also presented. These test exclude the effect of inconsistent and changes in mass distribution in the topside and it is also cannot be employed to closely understand the riser dynamics in the TLP overall system.

Other extreme cases including the effect of vertical seismic excitation on the TLP motion response also become the centre of many studies. Kawanishi et al. (1987) analyzed TLP motion with earthquake excitation. Base on the analysis, they recommend increasing the number spring constant to reduce the maximum relative displacement. They also found that between maximum absolute acceleration and the change of the ratio basically have no clear and consistent relationships. A good note that interesting to address in this paper is that this finding are all base on theory with many assumptions, a good experiment need to be done to test this theory.

The effect of this seismic excitation is further studied by Chandrasekaran and Gaurav (2008). They tested a triangular TLP under high waves and done the earthquake motion analysis. They found in the result that based on Kanai-Tajimi ground acceleration spectrum, tether tension showing a nonlinear variation. This variation is analytically higher than the tether tension set by the rules and regulation and call for examination for seismic safety purpose. They further verify using the numerical method that a much less response in shown by the TLP built on deeper water to the wave and earthquake forces. Their focus on triangular TLP, suggested that further studies need to be done to understand the response of other various geometrical type of TLP.

Extreme weather case such as the occurrence of Hurricane Lili force an thorough assessment for deep water structure especially in regard with the structural integrity. Morandi et al. (2004) provide a statistical analysis of the TLP tendon tension in the course of hurricane Lili. These data, that been taken based on Shell Brutus TLP. They found that the TLP behave well and the measured responses were less than the design limit. This favorable result is reached although hurricane Lili environmental condition approached 100-year value. There is no notable loss of tendon tension during the extreme environmental event, this suggest that the design can be improved economically.

Hurricane Katrina that struck in 2005 call for modification of future development of TLP in Gulf of Mexico. Murray et al. (2008) presented a new design procedure for TLP in the region of Gulf of Mexico, this new design must consider the new metocean criteria after the hurricane happened. To measure the response of TLP for this rare extreme condition a model test is set up on the scale of 1:92 in the basin. A worth notable result from the model test show poor agreement using uncoupled model, these lack of agreement including setdown in the extreme offsets, riser and tendons prediction, and also extreme response. In contrast the fully coupled simulation shows good consistencies with the model test data measured. A lot of work still needs to be done to improve the TLP responses in the extreme weather cases.

1.3 TLP tether system

Although many research focuses on the environmental loads acting on the TLP, all these environmental loads are natural, some of them are proven to be random, cannot be changed and must be understand and accepted as a challenge for the design of the TLP offshore. But one always can improve the structure and design of the TLP. It is obvious the most important structural part of the TLP is the tendon or the tether system. Over recent years many researchers work on the tether part of the TLP. They mainly focused on fatigue, reliability, stability, length, cost effective and hydrodynamic part of the TLP's tether.

One of many reasons a great effort been put to further increase the tether reliability is the fact that the high cost of in-service inspection. Wirsching et al. (1995) based on Heidrun TLP that been installed in the Norwegian sea done a maintainability analysis. They also included the fatigue and fracture reliability analysis for the TLP tether system that constantly exposed to the tensile load that is oscillating during the operation. This analysis claimed to quantify the performance of the structure and to check the structural integrity. They noticed that compared to fracture in course of extreme loading, fatigue remain as a dominant mode. The estimated reliability also enhanced notably when the assumption is made regarding the stress endurance limit.

Also based on the Heidrun TLP, Dalane (1998) done a fatigue reliability analysis. The study discussed the comparison of the numerical calculations and the forces acting on Heidrun tethers during the operational phase. He noted that to increase the fatigue reliability, the most effective way is to measure the fatigue loading. He further stated in the result that the cause of TLP resonant response for heave, roll and pitch periods is also because of first-order response that occur and not primarily caused by the second-order of loading. To better design the TLP, further studies need to be done to address this high frequency tether response.

Siddiqui and Suhail (2001) come up with the assessment of TLP tethers in term of reliability. This assessment used mainly fatigue reliability and fracture mechanics approaches. As the random loading from waves and wind constantly acting, the tether is exposed to fatigue failure. To estimate this damage Palmgren-Miner's rule is used while the estimation of reliability is done using Monte Carlo simulation and first order reliability method (FORM).

It is marked in the conclusion that the increase of water depth with a constant tether length shown a decreasing reliability index. They also noted that single tether joint is more reliable than a number of tether joints. While for the overall system the reliability index is marked at about 30%. These finding is very important especially for the designer in the early stage of TLP development, but as the industry grows and the exploration began to go deeper there is a need to further increase the reliability.

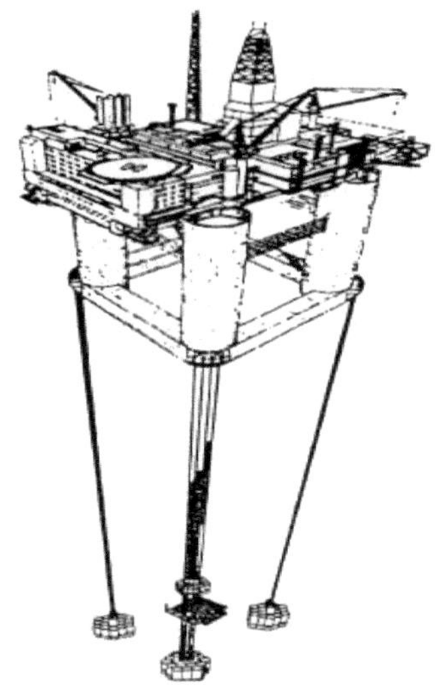

Figure: Heidrun Tension Leg Platform (Wirsching, 1995)

The needs to better understand the reliability of TLP tendons in shown by the emerging studies on this subject. Cicilia et al. (2008) presented design criterion for TLP tendons base on reliability. To generate the partial safety factors using the storm environmental situation in deep waters from Campeche Bay offshore Mexico, a long-term reliability-based method is applied. They found that a less scattered distinction of reliability indexes when the tendons designed using Load and Resistance Factor Design (LRFD). The result also observed the change of tendons thickness when the safety factors is calibrated. These findings serve as a good basis to further understand the methodology to assess the reliability, but other factors also needed to be address in order to establish a final safety factor for a specific TLP. Among of the factors include hull size, tendons material characteristic, updated environmental data and geometrical configuration of the TLP.

Besides the reliability part of the TLP tethers, the stability of the tethers also is a major concern from the design point of view. Chandrasekaran et al. (2005) used Mathieu stability analysis in order to define the value of vibration in the tethers. The numerical calculation proven the fact that to achieve the stability with increasing hydrodynamic loading, the tether tension needed to be increased. This increasing tether tension is also is a good help to gain platform stability. The study also noted that for the application of TLP in deep water, higher initial pretension is essential to improve the stability. This understanding provide a good basis in order to understand the relationship between increasing water depth and tether pretension.

The hydrodynamic part of the TLP tendon also one of the aspect that cannot been discarded when designing the TLP tendon. Wang and Zou (2006) discussed this subject and focused on several key hydrodynamic factor that include tendon springing and ringing response, vortex induced vibration (VIV) and tendon bottom tension slacking. De Boom et al. (1983) compared computational method based on potential theory and experiment in model basin to predict the tether motion and forces. They emphasize that during the early design phase the computational method are fairly applicable. They further noted that because of small underestimation of wave drift force using the computational method some inconsistencies also found. These inconsistencies need to be investigate further.

El-gamal and Essa (2013) conducted a study to observe the effect of tether length to the response of square TLP. The numerical calculations were executed in the time domain using modified Morrison equation with Airy's linear wave theory. The analysis however considers only the regular wave from surge direction. Based on the result they establish the relationship that is directly proportional between tether length and the natural time period. They also observe the decreasing tether stiffness when the tether tension is increased. This finding however needs to be further investigated considering all other wave propagating from other direction beside the surge direction.

1.4 TLP tendon material

The material aspect of the tendon is one of the big concerns in the development of TLP. Going deeper than 1000 meters made steel tendons unpractical, this is because of the fact that the weight is excessively high. Steel tendons also have a serious issue with elasticity that causes a resonance problem. Other factor that excludes the use of steel tendon in deep water is the cost effective issue. All these pushed the development of a new lightweight tendon.

One of the early notable and important studies on TLP tendon is made by Hanna et al. (1989), where they account the effect of several parameters including tension, weight, material and hydrostatic pressure on the TLP tendons. Lightweight materials are proposed to reduce tendon top tension, which have a lower tendon weight in air. Steel tendons can be used in deep water by increasing the ratio between diameter and wall thickness (D/t), but due to hydrostatic collapse in deep water, D/t ratio for steel tendons allowable only up to 15. For this reason new lightweight material are needed for deep water purpose, and their performance need to be clarify.

The performance of composite tendon compared to conventional steel tendon is studied by Wu (1999). This comparison shown some several superiority of composite tendon compared to traditional steel tendon, these includes significant reduction of horizontal offset, reduce pretension and maximum dynamic tension. The results of the TLP motion at center gravity are given in Table 3.

	TLP with steel tendon				TLP with composite tendon			
	Surge	Sway	Heave	Roll	Surge	Sway	Heave	Roll
Mean	236.98	235.49	-7.82	0.02	218.44	217.27	-5.74	0.00
Wave Frequency								
Tz	14.81	14.81	13.57	11.96	14.36	14.56	12.21	10.86
RMS	8.22	7.93	0.53	0.10	6.91	6.64	0.50	0.11
Max	26.98	26.54	1.79	0.49	23.62	23.11	1.60	0.46
Min	-26.85	-25.93	-2.5	-0.47	-22.72	-21.84	-2.05	-0.51
3hr Max	29.85	28.81	1.93	0.38	25.16	24.15	1.83	0.43
Low frequency								
Tz	169.11	153.74	188.12	67.3	182.96	154.60	195.42	77.71
RMS	18.36	17.51	1.18	0.02	18.24	17.41	1.00	0.01
Max	48.11	45.46	2.69	0.11	42.45	43.18	2.54	0.05
Min	-47.85	-45.03	-3.28	-0.05	-52.66	-47.85	-2.62	-0.12
3hr Max	52.93	51.06	3.36	0.07	52.09	50.74	2.84	0.03
High frequency (springing)								
Tz	4.67	4.62	4.04	3.9	4.96	4.89	4.47	4.23
RMS	0.05	0.05	0.08	0.01	0.08	0.08	0.13	0.02
Max	0.28	0.27	0.38	0.05	0.37	0.40	0.59	0.09
Min	-0.39	-0.31	-0.39	-0.05	-1.26	-1.61	-0.57	-0.08
Max Dynamic	69.65	67.20	4.81	0.47	66.28	64.42	4.43	0.53
Max Total	306.63	302.69	-12.62	0.50	284.73	281.69	-10.17	-0.54

Table 3 - Summary of TLP motion at center gravity (Wu, 1999)

As the steel tendons are viewed as unpractical and uneconomical, new materials have been developed to provide alternative solution. Botker et al. (2002) discussed the use of composite that are been developed by the Composite Alliance that consist of Kvaerner and Conoco in 1995. This composite material that consists of carbon fiber rod is set to be used for tether with 3000 meter water depth. They stated that composite tethers are better in term of stiffness to weight ratio and also strength to weight ratio. These composites have been proven to be competent through various testing including prototype testing.

The compatibility of carbon fiber composite to be used in the tether for TLP in ultra deep water is further clarified by Sparks et al. (2003). The new material that been discussed is the carbon fiber composites that have been developed by joint action of Freyssinet, Soficar, Bostik Findley, Institut Francais du Petrole and Doris Engineering. The carbon fiber tendon which consists of 19 rods is able to overcome the water pressure in deep water without damage, their also naturally buoyant and can be spool to be shipped to the field location. Other new technology in materials such as cellular tendons as proposed by Yu (2013) also provides a possible alternative for a more cost effective material in deep water. With all these structural capabilities of the tendons, a thorough analysis from economic point of view must be performed considering the material cost, fabrication and installation to establish its applicability.

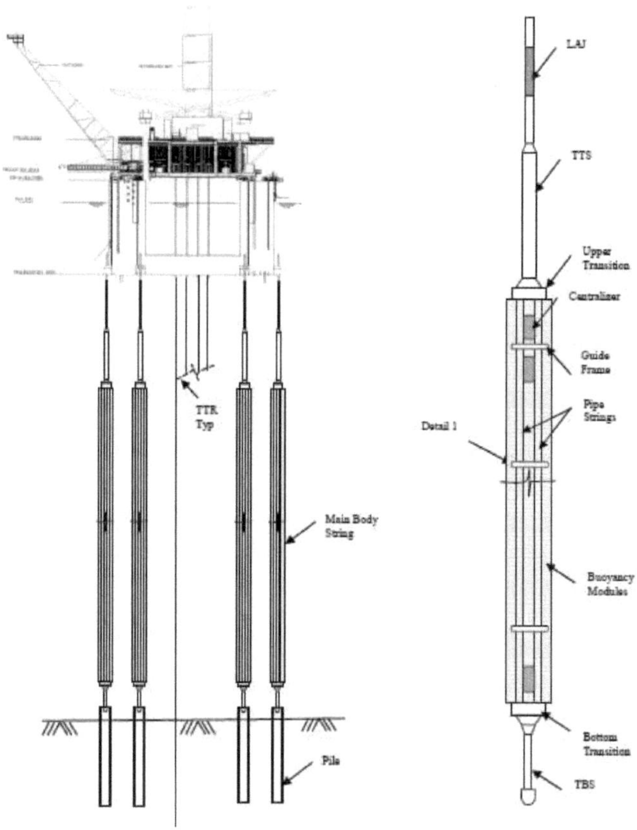

Figure: Cellular Tendon proposed by (Yu, 2013)

1.5 Analytical method and approach

Apart from the tendon or tether system, many researchers try to predict the behavior of the TLP under various type of environmental loading. Over the course of time, massive work have been done to analyze the TLP motion responses in different sea environment and using various method and approach. All these analyses proved that the importance of the TLP, as a superior choice for a production platform in the deepwater and ultra-deep water locations.

B.J. Natvig et al. (1992) presented TLP global motion performance analysis procedure. He recommended a number of analytical tools to perform the analyses. SEA-WAM is used to provide fast and accurate panel modeling. Then to generate panel pressures due to radiation and diffraction effects WAMIT is used. The slow varying wave drift induced motion that is very important to the global analyses is also computed by WAMIT. All six degrees of freedom rigid body motion response analysis is performed by SEAFLOATER.

Investigation of the coupled response of a TLP that is non-linear is presented by Ahmad and Ahmad (1998). The main objective of the investigation is to produce a control strategy of the force magnitude that help the TLP remain intact in severe ocean environment. To accurately quantify the optimal control force the Riccati equation is used. They imply that the control algorithm is less effective to the high frequency, this result is in contrast with low frequency range where it is more effective in the observation. They go further to suggest that base on the good result the lateral mooring may be replaced with response control. To avoid damaging TLP response in extreme cases active control is suggested to add more position-keeping ability. Although this control device proven helpful for station-keeping and response of the structure for some extent, but introducing this device may prompt a new pattern of fatigue failure due to the stress fluctuations that may occur.

In many cases prediction and assumption is made on calculation to simplify and accommodate the analysis process. Younis et al. (2000) designed a study using computational fluid dynamics (CFD) and design-code techniques to estimate the hydrodynamic loads on TLP. When the CFD result is compared with the DNV design recommended value, it is observed that drag coefficient for circular columns is in exact match. While this is not the case with square pontoon where the value for DNV result is higher than the CFD value. Further experiment data that were compared especially from wind-tunnel and towing-tank test strengthen the fact that it is fairly practical to use CFD for design application. These tools are seemed not only practical but provide more economical solution for TLP design and measured responses.

Global performance analysis also been done base on real existing TLP. S. Leverette et al. (2004) done an integrated global performance analysis of Matterhorn SeaStar TLP. The main goal of the analysis is to document the competence of the TLP design system to hold out against various environmental forces with regard to global parameters such as maximum tendon tension and maximum offset.

The effect of the wind loads to the TLP was computed using WINDOS, a program commercially developed by MARIN. To include the present of current loads the formulation provided in the API RP 2T is used. To address the forces that come from the waves, OSAWAVE is used to calculate the wave-frequency hydrodynamic loads using three-dimensional radiation/diffraction panel program. Model test is then conducted to validate the design methodology. From the model test it is observed steep waves that were in contact with the column result in high frequency tether loading. The study provide an excellent methodology especially to assess the response of a monocolumn TLP, and might be use to assess other type of TLP.

The analysis for response of TLP hull alone cannot be used as a basis for the overall design for TLP in deepwater. One must include the interaction of the hull with the tendon and riser system that result in the tension loads. Chen et al. (2005) provide an insight regarding the couple dynamic analysis for a mini TLP. The motions and tensions that needed for station-keeping are computed using a specific developed code, named COUPLE. The result from the code is then compared with the measurement taken from the experiment. They confirm that based on the comparisons the code can provide a solution to predict the dynamic interaction between tether and riser system with the TLP hull, but cannot accurately solve other quasi-static analyses. It is also shown that using Morison equation we can accurately predict the wave forces, this may be done only for the case of the wavelength is longer than the TLP columns and pontoons. To improve this limitation, a work needs to be done more so that motions and tensions can be predicted whether it is in low frequency or high frequency range.

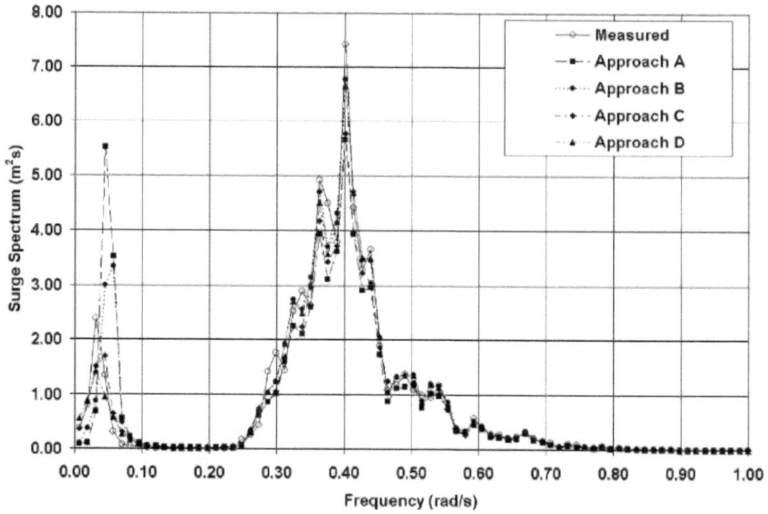

Figure: Surge response spectra for mini-TLP (Chen et al., 2005)

Tabeshpour et al. (2006) presented an analysis of the TLP dynamic using only numerical approach. This dynamic response in time domain is done using Modified Euler Method (MEM). The Pierson-Moskowitz spectrum is applied to form the time history of random waves, and Airy's linear wave theory is the based to determine the hydrodynamic forces. This analysis however doesn't contribute much to the existing analysis, a methodology to validate the numerical calculation is also lacking.

The dynamic analyses usually have some problem with the stiffness matrix, it is also has a concern about the inconsistency based on energy balance for this kind of matrix. To address this problem Senjanovic et al. (2013) derived a new stiffness matrix, the result of this calculations are later compared with FEM analysis. They noted that the restoring stiffness that inherit nonlinear behavior create a significant effect to the TLP dynamic response. These improvement especially for the time domain calculation might be a good reference to upgrade the Classification Rules, especially the one that involved construction and design part of the TLP.

Based on the review it is known that great work have been done so far to analyze the TLP. These analyses consider various environmental conditions and used different type of method and approach. These methods including using numerical calculation with the help of developed computer program such as WAMIT, SEAFLOATER and OSAWAVE. It is also known that using CFD is practical for this kind of analysis. Other method including model test in the basin and the measurement from real existing operating TLP. The review also strengthens the notion that using numerical calculation for the analysis, and validated the result with model test is a good methodology to present the motion response of TLP.

1.6 Motion response of tension leg platform

It is well known that the structure will respond to the random environmental loads. So far a thorough literature review has been made to discuss the works that have been done to understand the environmental loads and how to quantify them. The key components of TLP which is the tendon or tether itself also have been discussed. Various type and method of analysis that been done till this very time also been examined. All of these come to a point where researcher tries to answer, how the TLP will respond. Answering this question is very crucial for a safe design and operation of the TLP.

Winterstein et al. (1994) conducted a study about the springing and slow-drift responses of TLP. They try to predict extreme responses and fatigue, this prediction then is compared to the simulation result. Based on the TLP model, it is confirm that a mild non-Gaussian effect are observed on fatigue, this effect is true even during springing seastates. From practical point of view this study which highly theoretical, need a further proper testing and experimental prove. Deeper study need to be done in order to understand this type of response. The nature of sea waves is often stochastic, which mean in other word, having a random pattern. Adrezin and Benaroya (1999) made a study to predict the TLP response under this stochastic loading. They used a cylindrical body to manifest the hull and model the tether as massless springs. Borgman's method is then used to create the time history. They study the two cases, one with the hull mass two times smaller than the tendon's mass the second case with the hull mass two times bigger than the tendon's mass. It is observed that in the result the surge response increase significantly for the first case when the tendon forces are included. Such response however is not well shown for the second case. Because the tendons are modeled as massless spring the effect of weight are not well documented in regard of the response, this part need to be further analyze.

The response of the TLP is not only depending on the pattern of the loading, the geometrical aspect of the TLP also played a role. Chandrasekaran and Jain (2000) compared the dynamic behavior of triangular and square TLP under regular wave force. The non-linear behavior that come from the drag force and tether tension also comprised in the analysis. The response to the regular wave shown by the triangular TLP is very limited in the surge and heave degree of freedom compare to square TLP.To further confirm this result, a similar study need to be done so that this geometrical effect can be further confirm considering stochastic waves pattern. Among the response that have been seen so far from the TLP is not only the motion in surge, heave and sway, vortex induced vibration (VIV) also observed on TLP response. Laverette et al. (2003) analyze these VIV responses due to Eddy current in the Gulf of Mexico. The analysis also computes the stress responses that exist in the structure using FE analysis. For the deep water design purpose it is the coupled dynamic analysis especially for the full entire structure is seen very essential. If the development intends to go further deeper one must consider full coupled analysis for safe design.

TLP response with the effect of added mass are further studied by Tabeshpour et al. (2004) using analytical method. The solutions are calculated using first order pertubation method, and with non-harmonic Fourier expansion is used to solve the differential equation. These mathematical presentation help to provide an understanding of the conceptual view of a sea wave forces that trigger the heave response of the TLP. More detailed conceptual basis is needed to be established especially for the lateral response of the TLP. Ringing and springing response of triangular TLPs are studied by Chandrasekaran et al. (2011). The method to analyze the response is discussed and the mathematical formulation is developed for non-impact and impact waves. This ringing response must be handle properly in order to avoid severe breakdown of the platform. Based on the result they suggest the use of triangular TLP in deepwater because of its mild response especially in pitch degree of freedom. However this focused on TLPs with triangular and square geometry, and doesn't discuss other various types of TLPs such as monocolumn that maybe inherit a better motion response.

The non-linear behavior of the TLP response is presented by Abou-Rayan (2014). He tried to compare the difference of TLP geometrical configuration to the coupled response. These non-linearities that arise because of the hydrodynamic forces are analyzed using modified Morrison equation in time domain. The analysis is simplified using only uni-directional waves. He confirmed the mild response inherited by the triangular TLP. This analytical result is fairly consistent with Chandrasekaran et al. (2011) TLP dynamic behavior.

1.7 Model test with truncation method

One of the problems associate with the motion response of a structure moored in ultra deep water is the model test that needed to be done for validation. This is true primarily because of the limitation of the water depth in the test basin. With this depth it is also very difficult to scale down the model in a practical and rational manner.

A simple model test for TLP is presented by Kurian et al. (1993) , this model however did not study the ultra deep water situation. Motion of TLP in surge, sway and heave are studied using a model test to do the hydrodynamic analysis. The model was scaled down to 1:50. They neglected the forces that came from the current and wind. The axial stiffness of the tendons also been neglected in the modeling procedure. There is a need for a fresh study that includes these factors that previously been neglected especially with the new advance in technology in order to stay realistic.

To overcome this challenge associated with the ultra deep water various type of approach have been presented and proposed. Stansberg et al. (2002) discussed this problem and presented an overview of the hybrid approach. The hybrid approach comprises computer simulation and a model tests with truncated mooring lines. Active hybrid approach used actuators to artificially provide damping effect and mooring lines dynamic for the truncated parts of the tether. These actuators are controlled by computer in real time. With the limitation of test basin, hybrid approach seems more and more reliable especially with increasing water depth, the use of sale models that are too small will bring the problem that associated with practicality and accuracy. The most important question that needs to be further studied and clarified is how accurate this truncation method compared to full-depth model testing.

Choosing a right scale model also very important in order to clarify the uncertainties related to our previous question. Figure 3 shows how Stansberg et al. (2002) illustrate the dependency of verification on the model scale. One must always bear in mind when scaling the model test and truncate the line, there is an optimum scale need to be satisfied.

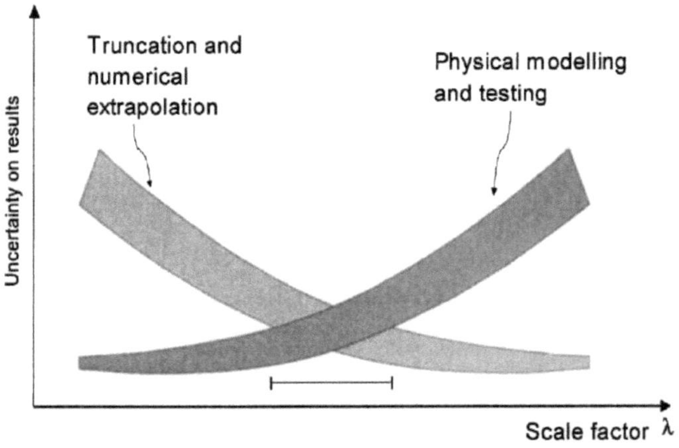

Figure 3 – Balance between uncertainties related to truncation and to small scales (i.e. large scale factor) (Stansberg, 2002)

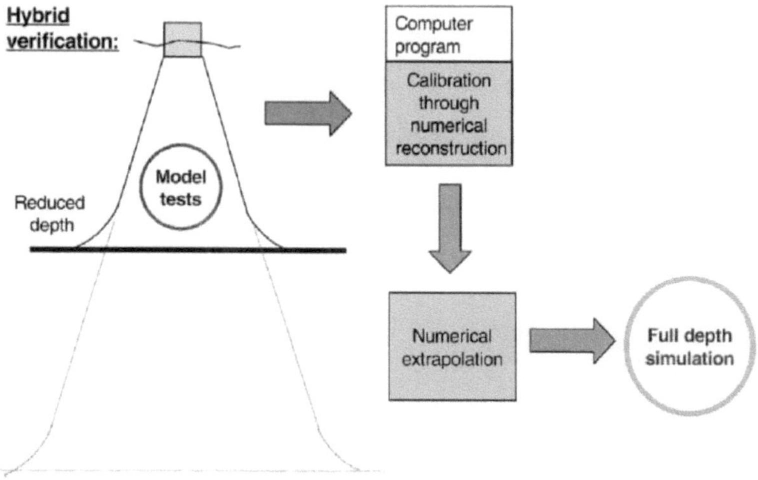

Figure: Hybrid verification procedure (Stansberg, 2002)

Qiao and Ou (2010) have done the dynamic analysis of semi-submersibles using numerical and model test with truncated mooring lines. Non-linear FEM is used in time domain to simulate the mooring lines dynamic, and to observe the differences between a full-depth and truncated mooring lines. They also considered the design of truncated mooring system and examine it difference with the full-depth mooring system as a major problems. Based on the result a good agreement is reach between the truncated mooring lines and full depth mooring lines except for a few aspects such as mooring line tensions in irregular waves, the RAO and damping.

Because of many limitations poses by the passive truncation method especially the difficulty to generate equivalent effect of the behavior of truncated mooring lines that mimic the behavior of the full-depth mooring line, active method is seen as the most reliable method. Cao and Tahchiev (2013) studied the active truncation mooring lines using actuator controlled by PID controller in the ocean basin. They confirmed the practicality of active truncation method but a more intelligent controller are needed, a kind that can adapt independently especially for simulation in random sea state.

1.8 Review of guideline, regulation and standard

Guidelines, rules, regulations and recommended practices from various bodies and class societies are also been reviewed in order to conduct the study according to existing updated standard.

A Joint Industry Project (JIP) conducted by Det Norske Veritas (1995) comes up with guidelines to perform structural reliability analysis for TLP. In the example, they analyze only 1 tether out of 16 using FREECOM-3D to generate transfer function that describes the tether response. They also include fatigue analysis for structural reliability as an important component.

In 2010 Det Norske Veritas (DNV) release a recommended practice for global performance analysis of deepwater floating structure. The recommended practice elaborate how to model the loads acting on the floater including hydrostatic, wave, wind, current and vortex-induced loads. The coupled and de-coupled response analyses also were discussed in details. The equations for all of these analyses also were presented in detail.

A few notes also worth mentioning based on the offshore standard release by DNV in 2011 on structural design of TLPs using Load and resistance factor design (LRFD) method. A time domain analysis is recommended especially for the global motion response, this is necessary in order to include the non-linear effect. For TLP in deep water, it is required to perform the fully coupled analysis that includes the interaction of the hull, tendons and risers. Finally to check the TLP design a model test needs to be done in the wave basin.

A set of rules from other classification society such as Bureau Veritas (BV) also been reviewed. This includes the rules for the classification of tension leg platform (2012) and classification of mooring system for permanent offshore units (2012). For the environmental design BV required 100 years return period for the extreme environment for the system to resist the loads and stay operable. For the case of survival environments 1000 return period are required.

In response to the random loading the maximum offset must be determine considering the wave response, low frequency response and also the mean offset. To analyze the TLP tendon, the static loads from tendon weight and buoyancy, offset caused by wind and current and also the tendon pre-tension must be included, the motion of the hull that contribute to the dynamic loads also need to be calculated.

The recommended procedures and guidelines provided by the International Towing Tank Conference (ITTC) regarding the truncation of model test and integration with the numerical simulations (2008) also been reviewed.

1.9 Conclusion

Based on detailed literature review in this chapter it is evident that the effect of weight reduction in the tendon or tether system on motion response of TLP if not few almost absent in the literature. With emerging technology in material and fabrication it seem very essential to conduct a research to find and test a lighter and stiffer tendon. This task is aimed to solve the problem of excessive weight of tendons in ultra-deep water. Solving this task will ultimately reduce the cost for tendons in ultra-deep water and also improve operational procedure such as transportation and installation.

Validation of the numerical result using a model test with truncation method also practically new. With the current trend of the industry to explore and produce hydrocarbon in deeper water and the limitation of water depth in test basin, the hybrid approach seem as the most practical and reliable way to test the structure motion for safe design.

Detail about method, steps and procedure of the research will be discussed in greater detail in the next chapter. The research methodology that will be applied for the research will be explained in detail in the next chapter.

REFERENCES

1. S. Chandrasekaran, A.K. Jain, and Anupam Gupta. *Influence of wave approach angle on TLP response.* Elsevier, 2007.
2. S. Chandrasekaran and Gaurav. *Offshore triangular tension leg platform earthquake motion analysis under distincly high sea waves.* Ships and Offshore Structures, 3:3, 173-184, 2008.
3. A. K. Jain. *Nonlinear coupled response of offshore tension leg platforms to regular wave forces.* 1995.
4. Toshimasa Kawanishi, Wataru Katoh and Hideo Furuta. *Tension leg platform earthquake motion analysis.* IEEE, 1987.
5. Yousun Li and Ahsan Kareem. *Stochastic response of tension leg platforms to wind and wave fields.* Journal of Wind Engineering and Industrial Aerodynamic, 36 (1990) 915-926, 1990.
6. Alberto Morandi, Diego Martinez, and Charles Smith. Statistics of TLP tendon tension behavior during hurricane Lili. *Proceedings of 23rd International Conference on Offshore Mechanics and Arctic Engineering 2004.* 20-25 June. Vancouver, Canada. OMAE2004-51445.
7. John Murray, Chan K. Yang, and Wooseuk Yang. *An extended Tension Leg Platform design for Post-Katrina Gulf of Mexico.* ISOPE-2008-287, 2008.

8. T.B. Johannessen, S. Haver, T. Bunnik and B. Buchner. *Extreme wave effects on deep water TLPs : Lessons learned from the Snorre A model tests.* DOT conference, Houston, 2006.
9. Amr R. El-gamal and Ashraf Essa. *Influence of tether legth in the response behavior of square tension leg platform in regular waves.* ISSN (Online): 2347-2820, Volume-1, Issue-1, 2013.
10. S. Chandrasekaran, N. R Chandak and Gupta Anupam. *Stability analysis of TLP tethers.* Ocean engineering 33 (2006) 471-482, Elsevier, 2006.
11. Jan Inge Dalane. Fatigue reliability – Measured response of the Heidrun TLP tethers. Marine structure 10 (1997) 611-682. Elsevier, 1998.
12. W.C de Boom, J.A. Pinkster, and S.G. Tan. *Motion and tether force prediction for a deepwater tension leg platform.* OTC 4487. Offshore technology conference, 1983.
13. N.A Siddiqui and S. Ahmad. *Reliability analysis against progressive failure of TLP tethers in extreme tension.* Reliability engineering and system safety 68 (2000) 195-205. Elsevier, 2000.
14. N.A. Siddiqui and Suhail Ahmad. *Fatigue and fracture reliability of TLP tethers under random loading.* Marine structure 14 (2001) 331-352. Elsevier, 2001.
15. E.D. Valenzuela. *Titanium stress joint for a TLP tendon system* . OTC 6910. Offshore technology conference, 1992.
16. Wang Tao and Zou Jun. *Hydrodynamics in deewater TLP tendon design.* Conference of global Chinese scholars on hydrodynamics, 2006.
17. Paul H. Wirsching, Demir I. Karsan and Shaddy Y. Hanna. Fatigue/fracture reliability and maintainability analysis of the Heidrun TLP tether system. *Proceedings of the ASME Offshore Mechanics and Arctic Engineering Conference*, Copenhagen, Denmark, June 19-22, 1995.
18. Stig Botker and Thomas B. Johannessen. *Composite risers and tethers: The future for deep water TLPs.* OTC 14176. Offshore technology conference, 2002.
19. Mohammed Y. Elasmai, Poojan Jhaveri, Harshvardhan Sindha, Carlo Vaccari, Dylan Barstow, and Ulysses Velasquez. *Santos basin tension leg platform utilizing embedded fiber optic in carbon fiber tendons*, Govenor's school of engineering and technology research journal, 2008.
20. Shaddy Y. Hanna, William H. Thomason, and Jerry G. Williams. *Influence of tension, weight and hydrostatic pressure on TLP tendons.* Journal of waterway, port, coastal and ocean engineering, Vol. 115, No 2, March 1989.
21. Shaddy Y. Hanna and Mamdouh m. Salama. *New tendon and riser technologies improve TLP competitiveness in ultra-deepwater.* OTC 12963, Offshore technology conference, 2001.
22. K.Z. Huang. *Composite TTR design for an ultra-deepwater TLP.* OTC 17159. Offshore technology conference, 2005.

23. C. Sparks, I.Zivanovic, J Luyckx and W.Hudson. *Carbon fiber composite tendons for deepwater tension leg platforms.* OTC 15164. Offshore technology conference, 2003.
24. S. Wu, m. Karayaka, and R. D'Souza. *Global performance comparison between composite and steel tendons in ultra-deep water TLPs.* OTC 11061. Offshore technology conference, 1999.
25. Jim Yu. *Cellular tendon- Enabling technology for ultra-deep water TLPs.* OTC 24305. Offshore technology conference, 2013.
26. S.K. Ahmad and S. Ahmad. *Active control of non-linearly coupled TLP response under wind and wave environments.* Computers and structures 72 (1999) 735-747. Elsevier, 1999.
27. Xiaohong Chen, Yu Ding, Jun Zhang, Pierre Liagre, John Niedzwecki, and Pier Teigen. *Coupled dynamic analysis of a mini TLP: Comparison with measurements.* Ocean engineering 33 (2006) 93-117. Elsevier, 2006.
28. Yang Jiahui and Ren Michael. *TLP hull structural design and analysis.* Engineering sciences Vol. 11 No. 4, Aug 2013.
29. Steve Leverette, Oriol Rijken, Mike Spillane, and Neil Williams. *Integrated global performance analysis of Matterhorn SeaStar TLP.* OTC 16609. Offshore technology conference, 2004.
30. Marco Masciola, Meyer Nahon, and Frederick Driscoll. *Preliminary assessment of the importance of platform-tendon coupling in a tension leg platform.* Journal of offshore mechanics and arctic engineering 153(3), 031901, 2013.
31. B.J. Natvig, H.S. Vogel and Oyvind Johnsen. *TLP global motion performance analysis procedures.* OTC 6889. Offshore technology conference, 1992.
32. I. Senjanovic, M. Tomic and S. Rudan. *Investigation of nonlinear restoring stiffness in dynamic analysis of tension leg platforms.* Engeneering structures 56 (2013) 117-125. Elsevier, 2013.
33. Mohammad Reza Tabeshpour, Ali Akbar Golafshani, and Mohammad Saeed Seif. *Efficient numerical dynamic analysis of tension leg platform under sea wave loads.* Journal of Marine Engineering, 2006.
34. B.A. Younis, P. Teigen, and V.P. Przulj. *Estimating the hydrodynamic forces on a mini TLP with computational fluid dynamics and design-code techniques.* Ocean engineering 28 (2001) 585-602. Elsevier, 2000.
35. Ashraf M. Abou-rayan. *Dynamic characteristics of offshore tension leg platforms under hydrodynamic forces.* International Journal of Civil Engineering (IJCE) Vol. 3 Issue 1, Jan 2014.
36. Ron Adrezin and Haym Benaroya. *Response of a tension leg platform to stochastic wave forces.* Probabilistic engineering mechanics 14 (1999) 3-17. Elsevier, 1998.
37. S. Chandrasekaran and A.K. Jain. *Dynamic behaviour of square and triangular offshore tension leg platforms under regular wave loads.* Ocean engineering 29 (2002) 279-313. Elsevier, 2002.

38. Chandrasekaran Srinivasan, Gaurav Gaurav, Giorgio serino and Salvatore Miranda. *Ringing and springing response of triangular TLPs*. International shipbuilding progress 58 (2011) 141-163. IOS Press, 2011.
39. Steve Leverette, Oriol Rijken , William Dooley and Hugh Thompson. *Analysis of TLP VIV responses to Eddy currents*. OTC 15289. Offshore technology conference , 2003.
40. Mohammad Reza Tabeshpour, Mohammad Saeed Seif and Ali Akbar Golafshani. *Vertical response of TLP with effect of added mass fluctuation*. XVI Symposium SORTA, 2004.
41. Steven R. Winterstein, Todd C. Ude and Gudmund Kleiven. *Springing and slow drift responses: Predicted extremes and fatigue vv. Simulation*. BOSS-94, Cambridge, July 1994.
42. Alex Argyros, Robin S. Langley and R.V Ahilan. Simplifying mooring analysis for deepwater system using truncation. *Proceedings of Twenty-first (2011) International offshore and polar engineering conference*. Maui, Hawaii, USA, June 19-24, 2011.
43. Rolf Baarholm, Carl Trygve Stansberg, Ivar Fylling, Ming-Yao Lee and Wei Ma. A robust procedure for ultra deepwater mooring testing. *Proceeding of Seventeenth (2007) International offshore and polar engineering conference.* Lisbon, Portugal, July 1-6, 2007.
44. Yusong Cao and Galin Tahchiev. A study on an active hybrid decomposed mooring system for model testing in ocean basin for offshore platforms. *Proceeding of the ASME 2013 32nd international conference on ocean, offshore and arctic engineering*. OMAE2013. June 9-14, Nantes, France.
45. V.J. Kurian, V.G Idichandy and C. Ganapathy. *Hydrodynamic response of tension-leg platforms – A model.* Experimental mechanics Volume 33, Issue 3, pp 212-217. September 1993.
46. R.S. Mercier, W.E. Schott, C.T. Howell, E.B. Denison, R. Gopalkrishnan and A.G.C. Ekvall. *Mars tension leg platform – Use of scale model testing in the global design.* OTC 8354. Offshore technology conference, 1997.
47. Dongsheng Qiao and Jinping Ou. Dynamic analysis of truncated mooring lines using numerical simulation and model tests. *Proceedings of the Twentieth (2010) International offshore and polar engineering conference*. Beijing, China, June 20-25, 2010.
48. Carl Trygve Stansberg, Harald Ormberg and Ola Oritsland. *Challenges in deep water experiments: Hybrid approach.* Vol. 124 ASME 2002.
49. Recommended practice for planning, designing, and constructing floating production systems. API RP 2FPS 2001.
50. Classification of mooring systems for permanent offshore units. Rule note NR 493 DT R02 E. Bureau Veritas, 2012.
51. Rules for the classification of tension leg platforms (TLP). Rule note NR 578 DT R00 E. Bureau Veritas, 2012.

52. Global performance analysis of deepwater floating structures. Recommended practice DNV-RP-F205. October 2010.
53. Structural design of TLPs (LRFD method). Offshore standard DNV-OS-C105. April 2011.
54. ITTC recommended procedures and guidelines. Testing and extrapolation methods, loads and responses, ocean engineering. Truncation of test models and integration with numerical simulations. 2008.
55. Technical report. Joint industry project. Guideline for offshore structural reliability analysis – Examples for tension leg platforms. Report No. 95-3198. Revision no. 02. Det Norske Veritas, 1995.